徜·徉

PROJECT TOUR
PTang Studio Ltd
编 著

大连理工大学出版社

图书在版编目(CIP)数据

徜徉 / PTang Studio Ltd编著. — 大连：大连理工大学出版社, 2013.12
ISBN 978-7-5611-8317-5

Ⅰ. ①徜… Ⅱ. ①P… Ⅲ. ①室内装饰设计—作品集—中国—现代 Ⅳ. ①TU238

中国版本图书馆CIP数据核字（2013）第256788号

出版发行：大连理工大学出版社
　　　　　（地址：大连市甘井子区软件园路80号　邮编：116023）
印　　刷：利丰雅高印刷（深圳）有限公司
幅面尺寸：240mm×300mm
印　　张：19.75
插　　页：4
出版时间：2013年12月第1版
印刷时间：2013年12月第1次印刷
出版策划：深圳市创福美图文化发展有限公司
责任编辑：初　蕾
责任校对：仲　仁
版式及装帧设计：DEHOMER
文字翻译：DEHOMER

ISBN 978-7-5611-8317-5
定　　价：320.00元

电　话：0411-84708842
传　真：0411-84701466
邮　购：0411-84703636
E-mail：designbookdutp@gmail.com
URL：http://www.dutp.cn
设计书店全国联销：www.designbook.cn

如有质量问题请联系出版中心：（0411）84709246　84709043

The exclusive distributorship in Taiwan is offered to ArchiHeart Corporation.
Any infringement shall be subject to penalties.

中国台湾地区独家经销权委任给ArchiHeart Corporation（心空间文化事业有限公司），侵权必究。

PTang Studio Ltd

PTang Studio Ltd is an architectural and interior design firm established in 1999. Having set up as a sole practitioner, founder and Principle Designer Philip Tang joined forces with business partner and Chief Designer Brian Ip in 2002. Since then, the Hong Kong China-based practice has expanded its focus from small residential projects to significant commercial and residential work across Asia, the USA and the Middle East.

Internationally renowned for its flexibility and modern design, the scale and type of projects PTang Studio Ltd undertake are diversified, ranging from residential and showflat designs to designs for corporate headquarters and movie launching functions.

Getting to basics, seeking for the finest quality and innovations, they develop a fresh and unique style that transcends existing boundaries and widens the horizon of design, yet create spaces that bring about the most comfortable atmosphere for clients by understanding their individual needs and preferences.

Their excellence can be reflected from the frequent appearances in various media, and rewards of different competitions.

The Studio is a five-time winner of the international Design Awards. It is also the winner of the Award for Global Excellence in Design from the International Interior Design Association – most recently as overall winner in the healthcare category.

PTang Studio Ltd 是一间成立于1999年的建筑及室内设计公司。在2002年，公司由邓子豪先生的个人设计业务发展至与首席设计师叶绍雄先生成为合作伙伴。从那时起，这间以中国香港为基地，以小型住宅项目为主的公司逐渐扩展至在亚洲、美国和中东等地设计著名的商业和住宅项目。

公司因灵活和现代化的设计而蜚声国际，承接项目的规模和类型广泛且多元，从住宅及样板间的设计，到企业总部及电影院的设计等。

公司从基础着手，寻求质感和创新，以清新及独特的风格，在超越现有的设计界限和扩大设计领域之余，亦不忘深入了解客户的个性化需求和喜好，为他们创建出最舒适的空间。

公司频频亮相于各种媒体，并在不同的比赛中屡获殊荣，这体现出了公司的优秀及卓越。

公司已五次荣获国际设计大奖。它也是国际室内设计师协会举办的全球室内设计比赛医疗保健类别的总冠军。

Philip Tang and Brian Ip   邓子豪及叶绍雄先生

NEO DERM THE CENTER	8
HILLSBOROUGH COURT	24
RESIDENCE 8 – CRYSTAL	34
LE BLEU DEUX	44
PARC PALAIS	60
SUKHOTHAI	74
SERENADE	88
MIAMI CRESECENT	108
LA ROSSA	130
SUNDERLAND	144
MOUNT EAST	158
BEL-AIR NO.8	170
DE YUCCA	182
MINGLY METRO	200
THE SAIL AT VICTORIA	230
NO.8 TAI TAM	244
RESIDENCE 8 – BLACK AND WHITE	266
GRAND PROMENADE	280
AWARDS LIST	296
ACKNOWLEDGEMENT	314

NEO DERM THE CENTER
Times Square | Hong Kong | China

"dynamically pure, shiny and bright"

Occupying a 3-storey unit in Times Square, the 65,000 sq ft flagship beauty center is amplified with a pure white curve staircase connecting all the floors. Special lighting effects designed along and under the steps captivate eyesight from all customers, making it the focal point of the center.

In the reception lobby, curvaceous contoured objects and designs are employed to create layers for the all white environment. The bright glittering white floor, coupling the white walls, visually heighten the low ceiling and contribute to an even more spacious and comfortable scene.

这间面积为6039平方米的三层美容中心旗舰店位于中国香港的时代广场，设计师以纯白色的弯曲楼梯连接不同楼层，空间更显宽敞。特殊的灯光设计沿台阶向下延伸，吸引了所有消费者的视线，使此处成为美容中心的焦点所在。

在接待大厅内，波浪状的曲线造型和设计为全白的空间营造了一种层次感。闪闪发光的白色地面，配合净白的墙壁，在视觉上提升了天花板的高度，使空间看起来更宽敞，也更舒适。

1/F

2/F

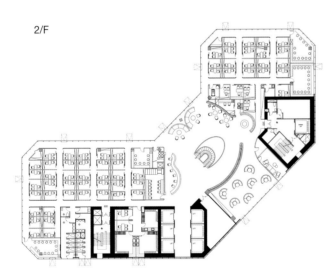

3/F

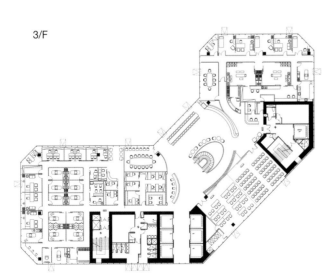

HILLSBOROUGH COURT
Central | Hong Kong | China

As the owner travels most of the time, not much storage space is required for the flat, leaving the designer plenty of flexibility when brainstorming design ideas. To further bring out the beauty of the flat's mountain view, designer uses white as the base colour, and green and wood as the accent colours. Having "natural" and "green" as the theme, "tree" has become a prominent element in the design.

The wall between living room and study room is removed and replaced by a glass window. While visually enlarged the area, showing the tree objects in white and green adds liveliness to the room. Reflecting under the lights, the shadow of the tree brings a casual alfresco feeling indoor. The bookcase, comprises with green and white storage boxes that consonant with the green boxes in the living room, unites the areas and at the same time, making it full of colours and layers.

In master bedroom, white wardrobe with laser-cut tree graphics and wooden side table offer a simple yet warm feeling, providing the best place for taking rest and relaxing.

因为户主经常出差，这间公寓并不需要太多的储物空间，所以设计师在思考设计构思的时候拥有更多的自由度。设计师选择了白色为主色调，并以绿色和木色作为强调色，配合窗外的翠绿山景，自然舒泰。设计的主题为"自然"和"绿色"，因此"树木"成为了这一设计中的主要元素。

设计师将客厅与书房之间的墙壁替换为玻璃窗。从视觉角度来说，白色和绿色的树枝物件在增加空间感的同时，也为房间带来了生机。在灯光下，树枝的阴影营造了一种户外的休闲气氛。书柜以白色和绿色的储物单元构成，与客厅的绿色储物柜相互呼应，使空间设计更为统一，并丰富了空间的色彩和层次。

在主卧室里，白色的衣柜配以激光切割的树木的图案，与木质的床头柜一起呈现出了一种简单而温暖的气氛，令主卧室成为了最佳的休息和休闲之所。

"simple, natural and green"

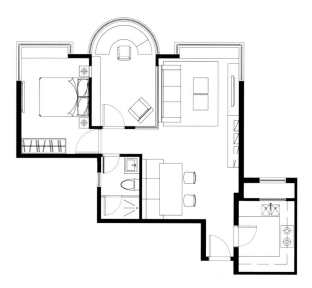

RESIDENCE 8–CRYSTAL
Dalian | China

"asymmetrical glitter"

With "crystal" as the theme, designer puts a big crystal feature on the wall right in the middle of the studio unit. Glittering in different angles, the feature furnishes the environment with a strong visual impact. Contrast to the white walls and furniture, the crystal feature stands out, radiates and lightens up one's spirit.

Wall, ceiling and furniture are all organic in forms and sleek in finishes. Asymmetrical yet neutral, dynamic yet peaceful, the design is in perfect balance.

设计师以"水晶"为设计主题,将一个大型的水晶体造型置于公寓单元墙壁的正中央。这一造型从不同角度来看都熠熠生辉,为室内空间营造了强烈的视觉效果。在与周围白色墙壁和家具的对比中,整个水晶体脱颖而出,使居者的心情轻松愉悦。

墙壁、天花板和家具都采用了有机形式和光洁的饰面。不对称中透露着素雅,动感中又隐含着宁静,这一设计完美地平衡了各种设计元素。

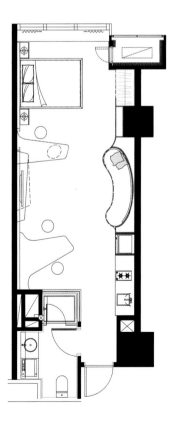

LE BLEU DEUX
Tung Chung | Hong Kong | China

"modern pop natural"

As the theme set by client, green and white are used as the main colour tones in the design, natural elements like leaves and trees can be found anywhere in the apartment. Usually designers would put a lot natural materials like bamboo, raw wood, and plants to achieve the feeling of "natural". However, in this apartment, you cannot find any of them, but a lot of tree and leaf graphics. Wallpaper, mirrors, feature walls and accessories containing elements of tree matched well with the modern design of the apartment.

本案的主题由客户设定,因此设计师使用了绿色和白色为主色调,而且在公寓中的任何地方都能发现树叶和树木等自然元素的踪迹。通常来说,设计师们会大量使用天然材料,比如,竹子、原木和植物等来营造"自然"的感觉。但是,在这个公寓里,你只能找到许多与树木和树叶相关的图形。包含这些图形的壁纸、镜子、主题墙和配件与公寓的现代设计风格完美融合。

PARC PALAIS
Kowloon | Hong Kong | China

"minimal details"

When entering the household, your eyes are stretched up towards the soaring height of the ceilings by the intricate 3d laser-cut feature on the main wall that is majestic in scale. The designers brainstorm white-on-white graphics to soften the extensive space. Otherwise, such a large expanse of plain wall, together with the huge glass windows, would most likely look stark and cold.

The design requirement is simple — to keep the home minimal, white and to bring out its sense of space. That is one of the greatest challenges as the space may look empty. Therefore, designers purposely utilise contemporary and colourful furnishings to jazz up the muted hues.

在进入此住宅之后，你的视线会被位于主墙上的复杂的3D激光切割造型引领至高耸的天花板。经过讨论，设计师们在白色墙壁上使用了白色的立体图形，这为诺大的空间增添了一份柔和感。否则，大片的白色墙壁，加上巨大的落地玻璃窗，会给空间带来一种僵硬和冰冷的感觉。

设计要求很简单，即以极简的白色装饰呈现出住宅的空间感。但这一简单的要求却成为了最大的挑战之一。在这诺大的空间里，如果不能成功使用极简设计，那么空间会有空无一物之感。因此，设计师刻意利用了色彩缤纷的现代配饰为这一素雅的空间增添一抹亮色。

1/F

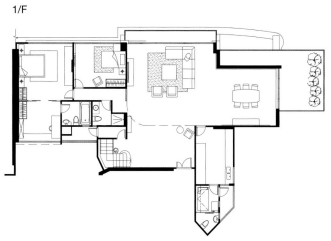

2/F

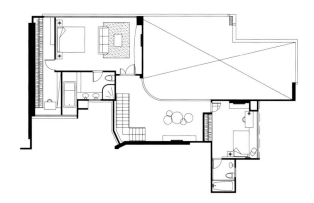

3/F

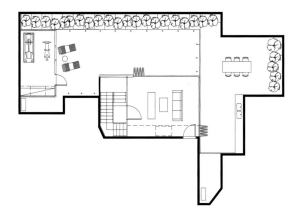

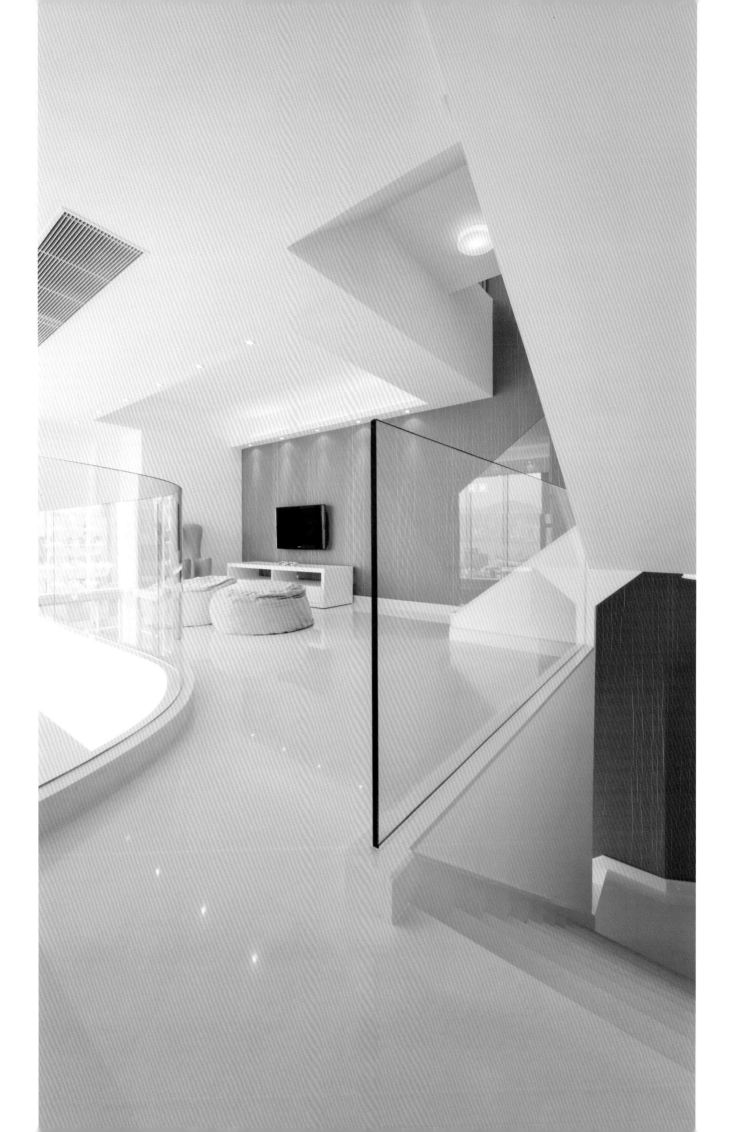

SUKHOTHAI
Bangkok | Thailand

"warm spacious box"

Designers ingeniously play up the double-height living space by connecting the levels with the same material from top to bottom.

To enrich the ample "boxlike" flat, the designer incorporates diverse style and texture of materials including wallpaper, wood, teak frame, granite, silk panel, glass, and adorns it with graphics and sculptures. Furnitures in different shapes and colours, yet with the same ambience, form a warm and comfortable relaxing area.

设计师通过从上至下使用相同材料将不同层面连接起来，巧妙地展现了双层高度的客厅空间。

为了充实这一宽敞的盒式公寓，设计师还采用了多种不同风格和纹理的材料，比如，墙纸、木材、柚木框架、花岗岩、裱丝板和玻璃等，并配以装饰图形和雕塑。虽然每件家具的形状和颜色各异，但是它们都拥有相同的氛围，把整个公寓打造成了一个温暖舒适的休闲空间。

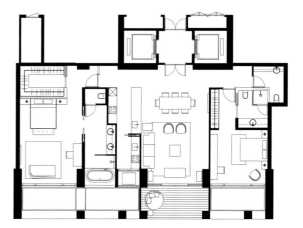

SERENADE
Mid-level East | Hong Kong | China

The scenery of Victoria Park and Victoria Harbour is highlighted by the sky terrace which is right in front of the main entrance of the apartment. While the sky terrace connects the separated living and dining areas, designer uses a mural painting along the corridor besides to unify the design of the whole flat. Carry on the design of the mural painting to another corridor, triangular leather panels of the same colour tone are bound together as a wall giving a 3-dimensional feeling. In master bedroom, designer replaces the original wall between the bedroom and closet area by a glass wardrobe. It visually enlarges the space and creates a modern contemporary atmosphere.

In each area, designer treats the walls differently to form a rich feeling. Wooden panels, leather panels, wallpaper, marbles and mirrors, each wall elevation is an artwork on its own. As every piece of material and accessory is specially selected by designer from around the world, you would feel like visiting a museum once enter into the apartment. This apartment is like a model and presentation of luxury living.

当业主置身于公寓主入口正前方的空中花园时，维多利亚公园及维多利亚港的景色可以尽收眼底。'空中花园连接了客厅和餐厅，为了使整间公寓的设计风格统一，设计师沿走廊布置了一幅壁画。这一设计理念在另一段走廊上得以延续，相同色调的三角形皮革面板组合起来，为墙面增添了立体感。在主卧室里，设计师拆除了原有的墙体，利用一个玻璃的衣柜将卧室与衣帽间分隔开来。这在视觉上放大了空间感，并营造了一种现代时尚的氛围。

为了丰富空间体验，设计师在每一个区域的墙体上采用了不用的样式和材料：木质面板、皮革面板、墙纸、大理石和镜子等，每一面墙都成为了一件艺术品。而且，每一种材料和饰品都是设计师专程从世界各地搜集回来的，因此当你进入这间公寓之后，就如在参观一间博物馆一样。它是精致生活的一个样板和一种体现。

MIAMI CRESECENT
Fanling | Hong Kong | China

As a house for a stylish couple with their son and daughter, many colours and fabrics are adopted for garnishment in each area.

In living area, the strong texture granite floor, together with the black and white walls, creates a cool and contrasting impact at first sight. In the black and white environment, the orange sofa unswervingly becomes the focal point of the area. To allow more sunlight to enter into the house, an open kitchen with clean and bright colour is designed. With a bar counter, it is just the perfect place for casual gathering.

Golden bar feature along the staircase connects ground floor to roof and leads the visitors to the bedrooms. A shocking pink glass door divides the bedroom and bathroom of the daughter's room. The zebra pattern glass in the master bedroom, not only serves as a glass door to the bathroom, but also an artwork alone. All the beautiful configurations softly emerge the vitality of the young.

这是为一对时尚夫妇与他们的儿子和女儿设计的住宅，因此每个区域的装饰元素选用了不同的颜色及材质。

在客厅中，硬质的大理石地面与黑白色的墙壁一起营造出了一种炫酷的感觉，给参观者留下了对比强烈的第一印象。橙色的沙发被放置于这种黑白色的氛围之中，自然成为了客厅的焦点所在。为了让室内空间获得更多的日光，设计师采用了干净明亮的开放式厨房设计，再配合上吧台，使这里成为了休闲聚会的完美之处。

沿楼梯设置的金色栏杆连接了地面与屋顶，并引领参观者进入卧室空间。惊艳的粉红色玻璃门分隔了女儿房的卧室和浴室。主卧室里的斑马纹图案的玻璃门，不仅是一扇浴室门，还是一件艺术品。所有的漂亮装饰都适度地展现出了一种青春的活力。

"the drama
of colours"

G/F

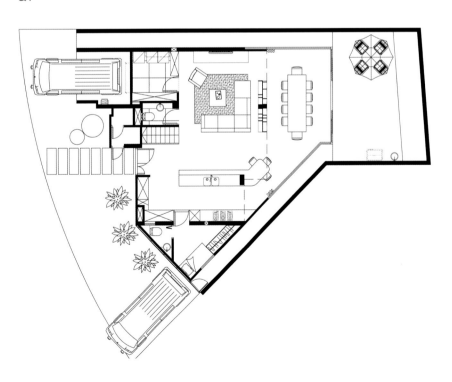

1/F

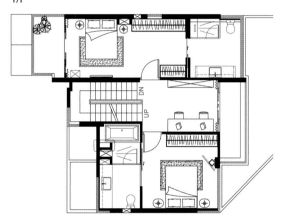

2/F

Roof

LA ROSSA
Tung Chung | Hong Kong | China

"red from the heart"

Stylish red and pure white are the skeleton of the design direction. Floral graphics around the home form a very warm atmosphere for this 3-bedroom sea view unit. A red TV cabinet acts like a red carpet, to serve as a foil to a contrasting tree wall graphic. This makes the square-like living room a romantic place.

A tree-like bookshelf in the study, is not only functional, but also becomes a focus of the unit. Its grey square boxes and the big red translucent flowers on the glass door create a dramatic contrast.

In the master bedroom, bedding, wallpaper and the wardrobe doors are covered by flowers. This continues the theme "flower" to the whole design.

设计师以时尚的红色和纯白色为设计的主调，让花样图案遍布家里的每个角落，为这间3个卧室的海景住宅营造了一种非常温暖的氛围。红色的电视柜就像红色的地毯一样，陪衬着墙面上的树形图案，使长方形的客厅变为一个浪漫的地方。

书房里的树形书架不仅发挥了自身的功能，而且成为了空间里的焦点。灰色方框与玻璃门上红色的半透明花朵相互辉映，营造了一种富含戏剧性的对比。

在主卧室里，床上用品、壁纸和衣柜门上也选用了不同的花形图案，使"花"的设计主题得以延续。

/ 135

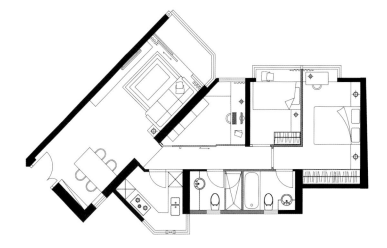

/ 137

SUNDERLAND
Kowloon Tong | Hong Kong | China

"brightly dark"

Located in the area of low-rise buildings, the flat enjoys sunlight coverage as much as it could during the day, where dark colour materials can be used in it with ease.

With metal, wood, wallpaper and leather situated in the same area, the long wavy carved pattern on the wall integrates all the elements comtemplating a harmonious ambience, and concurrently elongates the ceiling height for a roomy feel. The big bronze cupboard, made of the only metal element in the flat, comes hand in hand with the other furniture, yet characterises and finishes the area with a lustrous and cool taste.

In bedroom, an even darker colour tone is used to advance the contrast with the mirror walls, creating a cozy place for relaxation and rest.

这间公寓位于低层建筑中，在白天可以充分享受阳光的照射，所以即使选用较深颜色的设计材料也没有太大的问题。

墙壁上长长的波浪形的雕刻线条，与同一区域中的金属、木材、壁纸和皮革等材料融为一体，体现了一种和谐的美感，同时，在视觉上提升了天花板的高度，使空间更显宽敞明亮。大型的青铜色橱柜是公寓中唯一使用金属元素的地方，与其他家具完美融合，而且为空间带来了一种炫酷的感觉。

设计师在卧室使用了更深的色调，与镜面墙相互对比，打造了一处适合休闲的舒适场所。

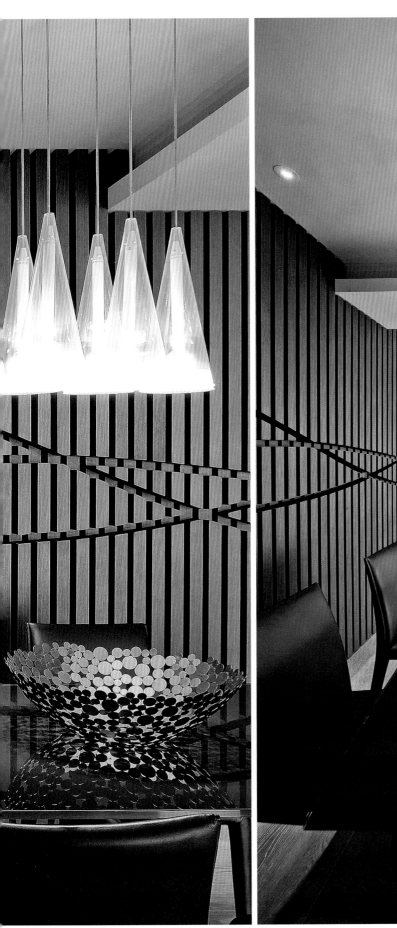

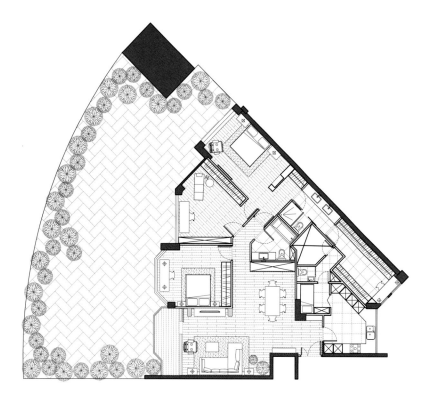

MOUNT EAST
North Point | Hong Kong | China

"multi-comfort"

Located in mid-level of North Point within a newly developed high-end residential area, this clubhouse accommodates the residences with a gathering and relaxing place for refreshment and leisure.

To bring about a contented and tranquil setting, the whole clubhouse uses earth tone colour scheme for a warm and natural feeling. The semi-transparent bronze curtains and crystal-like pendants from ceiling furnish a soft glimmering surprise.

Also using earth tone, the multi-function room introduces the most outstanding colour used in the clubhouse – the accent orange colour. It tenders a more energetic feel catered to all sort of activities.

The wave pattern wall, with continuity in the design of swimming pool, divides the outdoor area from the building.

这间会所位于北角半山一个新开发的高档住宅区内，作为住宅的配套设施，它为住户提供了一个可以放松身心和休闲聚会的地方。

为了营造一种令人满意的宁静的氛围，整个会所都采用了大地色系作为主要色调，温暖而自然。半透明的青铜色幕帘和如水晶一样的吊灯从天花板上垂坠下来，散发出柔和的光泽，给人以惊喜。

多功能厅也采用了大地色系，并以橙色作为点缀，成为会所中最闪亮的焦点。这种色彩可以迎合各种活动的需要，带给人一种充满能量和活力的感觉。

游泳池边的墙壁上装饰着波浪形的图案，它将建筑与室外空间分隔开来。

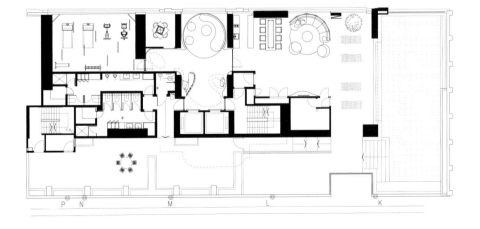

BEL-AIR NO.8
Island South | Hong Kong | China

"mix and match"

The designer applies different materials on different walls to make the apartment look rich and unusual. Once stepping into the apartment, a decorative copper strip wall panel is right at the entrance, allowing the owner to hang pictures and photos of different sizes on it in a random basis.

On the opposite side, artificial crocodile and horse leather are used to create a faux wall to conceal a storage cabinet for storing glasses and utensils of the owner. While creating a stylish wall finished with texture, it renders a nice and neat environment for the living room.

In master bedroom, there's a big floral pattern on the bed back wall. Contrasting with a metal curtain on the bedside, it creates a moody environment best for relaxing. Some of the design elements are extended to the design of the study room. While the geometry-shaped study table characterises the area, and the lighting effect in the bookshelves offers multiple compositions.

为了使这间公寓看起来既内涵丰富又与众不同，设计师在不同墙面上应用了不同的装饰材料。在步入公寓的入口之后，首先映入眼帘的是一面古铜色的条状的装饰墙，业主可以随意地在上面悬挂不同尺寸的图画和照片。

而在对面的装饰墙上，设计师采用了人造的鳄鱼皮和马毛，其后面隐藏着一个储物柜，是业主用来存放各式玻璃制品和器皿的地方。这种设计手法在打造了一面富含纹理的时尚的墙体的同时，也为客厅提供了一种干净整洁的环境。

主卧的背景墙上有一个大型的花朵图案，它与床侧的金属幕帘形成了对比，营造了一种惬意的休闲氛围。书房的设计沿用了一些相同的设计元素。几何形状的书桌赋予了这片区域以自身的特色，而书架的灯光设计也令空间更富有层次感。

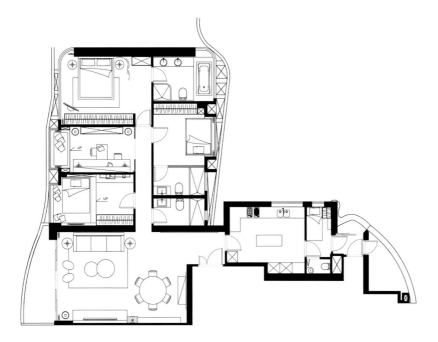

DE YUCCA
Shatin | Hong Kong | China

"modern classy"

High above the city of Shatin, the flat can enjoy 180 degrees of shimmering seascape. Connecting to private balcony and swimming pool, the sitting area which is mostly used, calls for a soft and comfortable feeling. In view of that, designers use different kinds of leather and fabric to create and augment an enjoyable warm feeling. A few classic elements like picture frames and mirrors are incorporated into the design for a more contemporary touch.

In bedrooms, wallpapers and drapes that boast somewhat traditional motifs, such as floral pattern, are put to use. Combining furnishings and accessories that are distinctly sleek and streamlined in silhouette, it brings comfort and peace in another way.

这间公寓坐落于沙田半山上,业主可以观赏到180度的迷人海景。起居室连接了私人阳台和游泳池,这里是最常使用的地方,因此业主要求拥有一种舒适的感觉。考虑到这一点,设计师采用了不同种类的皮革和织物,营造了愉悦的和温暖的氛围。相框和镜子等经典元素也在公寓中有所体现,增添了一份现代感。

在卧室中,壁纸和幕帘上体现的都是传统图案,如花卉图案等,它们与光洁的流线形的配饰一起,以另一种方式提供了一种安逸平和的感觉。

G/F

1/F

Roof

MINGLY METRO
Haining | China

"modest comfort, daring soul"

With its own garden, the architectonic detached house is designed with genuineness, comfort and simplicity. The use of different materials in each area does not only pronounce the feature of each area, but also link them up effortlessly.

Right next to the homey living room with raw tile stone and furry carpet is a panarama of wine collection. The square patterned black stainless steel wine cabinet exhibits the wine collection bottle by bottle neatly and spectacularly. Up along the staircase with glass partition, the sitting area and dining area on the second floor in light beige tone are illuminated by nature sunlight. Further up the stairs, it leads to the master bedroom with wooden floor and furniture, wallpaper feature wall, stainless steel partition and marble restroom. All the different materials mingle and create a harmonious environment.

这栋独立式住宅拥有私家花园，设计师为其营造了简单质朴、温馨舒适的氛围。每个区域采用了不同种类的装修材料，既突显了每个区域的自身特点，又非常自然地将各个区域连接了起来。

客厅中使用了未经雕琢的石材和毛绒地毯，给人一种温馨的感觉。客厅对面是以黑色不锈钢制成的方格状的酒柜，酒瓶在里面整齐壮观地摆放着，这是业主酒品收藏的一个全景展现。沿着带有玻璃隔断的楼梯拾级而上可以进入位于二层的起居室和餐厅，这里采用的是浅米色的色调并以自然光线照明。沿着楼梯继续向上可以进入主卧室，这里采用了木地板、木家具、墙纸背景墙、不锈钢隔断和大理石装饰的卫生间。所有材料完美融合，打造了一个轻松和谐的环境。

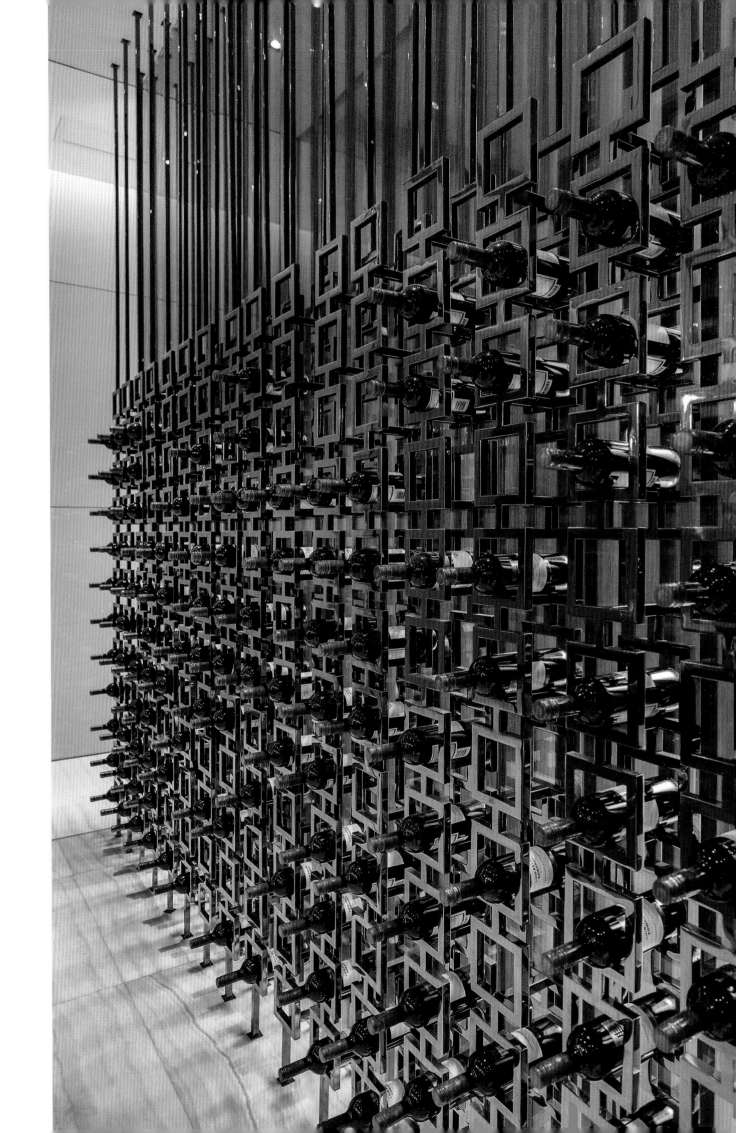

1/F

2/F

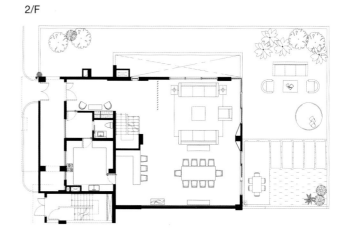

3/F

THE SAIL AT VICTORIA
Island West | Hong Kong | China

"oblique nature"

With "tree" as the design theme, but in order to avoid using green colour in the apartment, designers put a lot of visual elements of nature into the framing of the furniture, such as dining table, chairs, lamp, feature wall and shelves etc. in the form of "plants".

Strong distinct colours and graphics are imbedded in the apartment to create contrast. As the apartment is tiny, designers change the bedroom wall into a glass partition and glass wardrobe that looks slimmer. All furniture in the flat is tailor-made to fit into the environment and to maximize the usage of the space.

本公寓的设计主题是"树木",但是为了避免大量使用绿色,设计师在家具的框架中融入了大量的自然的视觉元素,比如,在餐桌、椅子、灯具、背景墙和架子上等,都出现了植物形态的装饰元素。

鲜艳的颜色与装饰图形在公寓中营造了强烈的对比感。因为空间狭小,所以设计师将卧室中的墙体改造成了玻璃隔断和衣柜,增强了房间的空间感。公寓中所有的家具都是为了这个空间而专门设计制作的,以使空间利用率实现最大化。

/ 241

NO.8 TAI TAM
Island South | Hong Kong | China

"contemporary classic"

Taking into account the location and the structure of the house – each floor has flat and straight lines and a relatively low ceiling, designer suggests a modern design style which incoporates classic elements with contemporary touches. To harmonize the many black walls requested by owner where he feels black is classy in feeling, designer embraces wallpapers and curtains with floral patterns that boast somewhat traditional motifs, coupled with beautiful streamlined furniture and furnishing accessories, to create two different styles at the same time.

The most distinguished feature of the house is made the most of by the designer. The private infinity pool located at basement doesn't actually get much sunlight at certain times of the day. To alleviate this problem, designer punches a hole through the floor of the spacious terrace above and paves it with glass. While conferring enormous sunlight to the pool, it engenders a spectacular view for the terrace.

考虑到住宅所处的位置，以及住宅平直的楼面布局和相对低矮的天花板等结构特点，设计师建议选用现代设计风格，即将经典元素与现代的质感融合起来。业主偏爱黑色，所以为了协调许多的黑色墙面，设计师使用了带有传统花朵图案的墙纸和幕帘，配以线条优美的家具及配饰，同时打造出了两种不同的风格。

住宅内最具特点的一处设计要归功于设计师。位于地下室的私人无边泳池获得太阳照射的机会并不多。为了解决这个问题，设计师在宽敞的上层露台的地面上打了一个洞并以玻璃覆盖，这样做既可以使大量的太阳光直接照射到泳池，也为露台提供了有趣的景观。

Roof

1/F

2/F

3/F

4/F

5/F

RESIDENCE 8 — BLACK AND WHITE
Dalian | China

Line graphics can be found in every corner of this flat. Lines are put into play on glasses, cabinets, carpets and walls. The lines not only visually decorate the interior, but also functionally hide the joints of the cabinets. Some of them are even used as handles of the cabinets.

To make the outcome of the whole design more dramatic, great contrast is in used. Black marble with strong texture on floor gives a strong impact in contrast with the high glossy white wall and cabinet.

A flexible working desk in the bedroom can be turned into a decorative cupboard, and replaced with a folding bed hidden in the cabinet, providing an appealing spectacle of home office.

在本公寓内的每个地方都可以发现线条图形，比如，玻璃上、壁橱上、地毯上和墙壁上等。线条图形不仅在视觉上点缀了室内空间，而且在功能上起到了隐藏壁橱接缝的作用，有些线条图形甚至直接被用为壁橱的把手。

为了使整个设计更具戏剧性，设计师使用了对比的手法。质感强烈的黑色大理石地面，与光洁的白色墙壁和壁橱相映成趣，给观者留下了深刻的印象。

在卧室内有一个灵活的工作台，可以变身为一处具有装饰功能的壁橱，也可以被隐藏在壁橱内的折叠床所取代，这一点体现出了居家办公的独特吸引力。

"all in black"

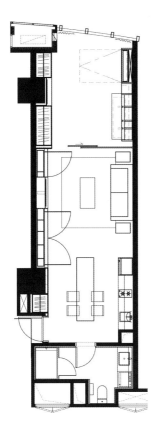

GRAND PROMENADE
Island East | Hong Kong | China

"romantic black"

The designer decides to go for an all-black look for the three-bedroom apartment which is quite a daring concept. Entering the home, your eyes immediately adjust to the dramatic space where a white curvy sculpture is accentuated by spotlights on a black mirrored cabinet.

Black walls contrast starkly against the tan wooden floor, the deep grains of which almost map out your path down the hallway and into the living area of the home. Another unique hallmark brainstormed by the designer is the feature wall of photographs that faces to the dining area. He has also taken pains to balance the mix of simple, streamlined furnishings with funkier items.

设计师决定为这间带有三个卧室的公寓选择全黑的设计，这是一种非常大胆的设计理念。一进入公寓，立刻映入眼帘的是一个极富戏剧性的空间，放置在黑色镜面壁橱上的白色雕塑经过聚光灯的照射更加突显了曲线的美感。

黑色的墙壁与棕褐色的木质地板形成了鲜明的对比。地板上清晰的纹理引导你穿过玄关进入到公寓的起居空间。餐厅对面的照片背景墙是空间中的另一个独特的设计标志。设计师努力地在简单的流线形的家具与时髦的配饰之间找寻一个平衡点。

/ 287

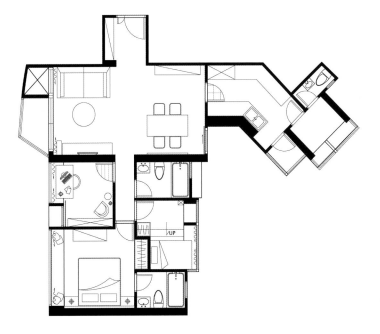

AWARDS THAT WE HAVE BEEN RECEIVED IN THE PAST FEW YEARS

奖项

Outstanding Greater China Design Awards 2012
Neo Derm The Center - Hong Kong China

2012 大中华杰出设计大奖
中国香港 Neo Derm The Center

International Design Awards 2012
Silver Prize
Mingly Metro - Haining, China

2012 国际设计大奖
银奖
中国海宁 Mingly Metro

IAI Awards 2012
Honorable Mention
Miami Crescent - Hong Kong China

2012 亚太设计双年奖
荣誉奖
中国香港迈尔豪园

IAI Awards 2012
Honorable Mention
Neo Derm The Center - Hong Kong China

2012 亚太设计双年奖
荣誉奖
中国香港 Neo Derm The Center

Asia Interior Design Award 2012
Apartment Unit - Silver Prize
Sukhothai Residence - Thailand

2012 AIDIA 亚洲室内设计奖
最佳公寓设计 - 银奖
泰国 Sukhothai Residence

Asia Interior Design Award 2012
Prototype Unit - Silver Prize
Residence 8 - Dalian, China

2012 AIDIA 亚洲室内设计奖
最佳样板房设计 - 银奖
中国大连 Residence 8

Asia Interior Design Award 2012
Non Residential Space Unit - Bronze Prize
Neo Derm The Center - Hong Kong China

2012 AIDIA 亚洲室内设计奖
最佳非住宅空间设计 - 铜奖
中国香港 Neo Derm The Center

Asia Interior Design Award 2012
Villa Unit - Bronze Prize
Miami Crescent - Hong Kong China

2012 AIDIA 亚洲室内设计奖
最佳别墅设计 - 铜奖
中国香港迈尔豪园

Awards 2012

IIDA Global Excellence Awards 2012
Best Interior Awards
Sukhothai Residence - Thailand
2012 国际室内设计协会设计大奖
最佳室内设计奖
泰国 Sukhothai Residence

**Andrew Martin
Interior Design Awards 2012**
Top 50 International Interior Designers
Philip C.H. Tang & Brian S.H. Ip
2012 安德鲁·马丁室内设计大奖
全球50佳室内设计师奖
邓子豪先生及叶绍雄先生

Perspective Awards 2012
Commercial, Retail or Office
- Certificate of Excellence
Neo Derm The Center - Hong Kong China
2012 透视设计大奖
商业、零售或办公空间 - 优秀奖
中国香港 Neo Derm The Center

**Asia Pacific
Interior Design Awards 2012**
Sample Space - Excellence Award
Residence 8-Crystal - Dalian, China
2012 亚太室内设计大奖
样板房设计 - 优秀奖
中国大连 Residence 8-Crystal

Asia Pacific Property Awards 2012
Private Residence
- Best Interior Design in Asia Pacific
Miami Crescent - Hong Kong China
2012 亚太地产奖
私人住宅设计 - 最佳亚太区内设计
中国香港迈尔豪园

Asia Pacific Property Awards 2012
Private Residence - Best Interior Design
Miami Crescent - Hong Kong China
2012 亚太地产奖
私人住宅设计 - 最佳室内设计
中国香港迈尔豪园

Asia Pacific Property Awards 2012
Apartment - Highly Commanded Interior Design
Sukhothai Residence - Thailand
2012 亚太地产奖
公寓设计 - 高度推荐奖
泰国 Sukhothai Residence

**Awards
2012**

International Design Awards 2011
Office Category - Gold
Neo Derm The Center - Hong Kong China
2011 国际设计大奖
办公室设计 - 金奖
中国香港 Neo Derm The Center

International Design Awards 2011
Commercial Category - Bronze
Mount East Club House - Hong Kong China
2011 国际设计大奖
商业空间设计 - 铜奖
中国香港明园西街会所

International Design Awards 2011
Residential Category - Bronze
Residence 8-Black and White
- Dalian, China
2011 国际设计大奖
住宅设计 - 铜奖
中国大连 Residence 8-Black and White

International Design Awards 2011
Residential Category
- Honorable Mention
Residence 8-Tree - Dalian, China
2011 国际设计大奖
住宅设计 - 荣誉奖
中国大连 Residence 8-Tree

International Space
Design Award 2011 - Idea Tops
Best Design Finalist Award of Clubs
Mount East Club House - Hong Kong China
2011 国际空间设计大奖 -- 艾特奖
最佳会所设计
中国香港明园西街会所

International Space
Design Award 2011 - Idea Tops
Best Design Finalist Award of
Residential Space
Hereford Road Residence
- Hong Kong China
2011 国际空间设计大奖 -- 艾特奖
最佳住宅设计
中国香港禧福道住宅

The 9th Modern Decoration
International Media Prize
Annual Show Flat Award
Hill Paramount - Hong Kong China
第9届现代装饰国际传媒奖
年度最佳样板房
中国香港名家汇样板房

IIDA Global Excellence Awards 2011
Best of Category - Healthcare
Neo Derm The Center - Hong Kong China
2011 国际室内设计协会设计大奖
保健类最佳设计
中国香港 Neo Derm The Center

Awards
2011

ApIDA

Asia Pacific Interior Design Awards 2011
Public Space - Excellence Award
Neo Derm The Center - Hong Kong China
2011 亚太室内设计大奖
公共空间设计 - 优秀奖
中国香港 Neo Derm The Center

design et al

International Design and Achitecture Awards 2011
Best Residential £20 million plus
No.8 Tai Tam - Hong Kong China
2011 国际设计及建筑奖
最佳住宅
中国香港大潭道8号住宅

Asia Interior Design Award 2011
Best Residential Design Company
PTang Studio Ltd
2011 AIDIA 亚洲室内设计奖
最佳居住空间设计公司
天豪设计有限公司

Asia Interior Design Award 2011
Category of Villa - Golden Prize
Hereford Road Residence - Hong Kong China
2011 AIDIA 亚洲室内设计奖
最佳别墅设计 - 金奖
中国香港禧福道住宅

Asia Interior Design Award 2011
Category of Apartment Unit - Silver Prize
Bel-Air No.8 - Hong Kong China
2011 AIDIA 亚洲室内设计奖
最佳住宅设计 - 银奖
中国香港贝沙湾8号住宅

Asia Interior Design Award 2011
Category of Prototype Unit - Silver Prize
Serenade - Hong Kong China
2011 AIDIA 亚洲室内设计奖
最佳样板房设计 - 银奖
中国香港上林样板房

Asia Interior Design Award 2011
Category of Apartment Unit - Bronze Prize
Hillborough Court - Hong Kong China
2011 AIDIA 亚洲室内设计奖
最佳住宅设计 - 铜奖
中国香港旧山顶道住宅

International Design Awards 2011
Residential Space / Single Level - Bronze
Serenade - Hong Kong China
2011 国际设计大奖
住宅设计（单层）- 铜奖
中国香港上林样板房

Awards 2011

International Design Awards 2011
Residences - Honorable Mention
Hillborough Court - Hong Kong China

2011 国际设计大奖
住宅设计 - 荣誉奖
中国香港旧山顶道住宅

Asia Pacific Property Awards 2011
Best Interior Hong Kong China
- Highly Commanded
Hill Paramount - Hong Kong China

2011 亚太地产奖
中国香港最佳室内设计 - 高度推荐奖
中国香港名家汇样板房

Asia Pacific Property Awards 2011
Best Interior Hong Kong China
- Highly Commanded
Serenade - Hong Kong China

2011 亚太地产奖
中国香港最佳室内设计 - 高度推荐奖
中国香港上林样板房

Awards 2011

Residence Award 2010
The 10 Most Influential Chinese Designers Award
Philip C.H.Tang
2010 精品家居大奖
最具影响力的10大华人设计师
邓子豪先生

IIDA Global Excellence Awards 2010
Residences - Honorable Mention
Hillsborough Court - Hong Kong China
2010 国际室内设计协会设计大奖
住宅设计 - 荣誉奖
中国香港旧山顶道住宅

Taiwan Interior Design Award 2010
Excellence in Residential Space / Single Level
Serenade - Hong Kong China
2010 中国台湾室内设计奖
最佳居住空间（单层）
中国香港上林样板房

 IAI **IAI**

Taiwan Interior Design Award 2010
Excellence in Residential Space / Single Level
Hereford Road Residence - Hong Kong China
2010 中国台湾室内设计奖
最佳居住空间（单层）
中国香港禧福道住宅

IAI Awards 2010
Residential Space Design - Silver Award
Serenade - Hong Kong China
2010 亚太设计双年奖
居住空间设计 - 银奖
中国香港上林样板房

IAI Awards 2010
Residential Space Design - Excellence Award
Bel-Air No.8 - Hong Kong China
2010 亚太设计双年奖
居住空间设计 - 优秀奖
中国香港贝沙湾8号住宅

IAI **IAI**

IAI Awards 2010
Residential Space Design - Excellence Award
Hillsborough Court - Hong Kong China
2010 亚太设计双年奖
居住空间设计 - 优秀奖
中国香港旧山顶道住宅

IAI Awards 2010
Residential Space Design - Excellence Award
No.8 Tai Tam - Hong Kong China
2010 亚太设计双年奖
居住空间设计 - 优秀奖
中国香港大潭道8号住宅

Awards 2010

IAI

IAI Awards 2010
Residential Space Design
- Excellence Award
Chelsea Residence - Shanghai, China

2010 亚太设计双年奖
居住空间设计 - 优秀奖
中国上海丁香御苑样板房

IF Design China Awards 2010
Commercial Space Design
- Excellence Award
Bel-Air No.8 - Hong Kong China

2010 IF 中国设计大奖
商业空间设计 - 优秀奖
中国香港贝沙湾8号住宅

Outstanding Greater China Design Awards 2010
Spatial Designer - Winner
Serenade - Hong Kong China

2010 大中华杰出设计大奖
空间设计奖
中国香港上林样板房

Asia Interior Design Award 2010
Best Residential Building Unit - Silver Medal
Grand Promenade - Hong Kong China

2010 AIDIA 亚洲室内设计奖
最佳住宅设计 - 银奖
中国香港嘉亨湾

Asia Interior Design Award 2010
Best Villa - Bronze Medal
No.8 Tai Tam - Hong Kong China

2010 AIDIA 亚洲室内设计奖
最佳别墅设计 - 铜奖
中国香港大潭道8号住宅

Asia Interior Design Award 2010
Best Residential Building Unit
- Excellence Award
16A Le Bleu Deux Showflat
- Hong Kong China

2010 AIDIA 亚洲室内设计奖
最佳住宅设计 - 优秀奖
中国香港水蓝天岸16A样板房

Asia Interior Design Award 2010
Best Residential Building Unit
- Excellence Award
19A Le Bleu Deux Showflat
- Hong Kong China

2010 AIDIA 亚洲室内设计奖
最佳住宅设计 - 优秀奖
中国香港水蓝天岸19A样板房

Asia Interior Design Award 2010
Best Residential Building Unit
- Excellence Award
The Sail at Victoria - Hong Kong China

2010 AIDIA 亚洲室内设计奖
最佳住宅设计 - 优秀奖
中国香港傲翔湾畔样板房

Awards 2010

Asia Interior Design Award 2010
Best Apartment Unit - Excellence Award
Parc Palais - Hong Kong China

2010 AIDIA 亚洲室内设计奖
最佳公寓设计 - 优秀奖
中国香港君颐峰住宅

Asia Pacific Property Awards 2010
Best Interior Hong Kong China - 5 Star
The Sail at Victoria - Hong Kong China

2010 亚太地产奖
中国香港最佳室内设计 - 五星大奖
中国香港傲翔湾畔样板房

Asia Pacific Property Awards 2010
Best Interior Hong Kong China
- Highly Commanded
No.8 Tai Tam Road - Hong Kong China

2010 亚太地产奖
中国香港最佳室内设计 - 高度推荐奖
中国香港大潭道8号住宅

ADEX - Design Journal 2010
Award for Design Excellence
Interior Design - Platinum Award
19A Le Bleu Deux Showflat
- Hong Kong China

2010 ADEX 国际设计大奖
室内设计 - 白金奖
中国香港水蓝天岸19A样板房

CIID (Golden Beach) China
Interior Design Awards 2010
Excellence Residential Design Award
No.8 Tai Tam - Hong Kong China

2010 中国国际建筑及
室内设计节"金外滩"奖
最佳住宅空间奖
中国香港大潭道8号住宅

Awards
2010

International Design Awards 2009
Residential Interior Design - 1st Place
16A Le Bleu Deux Showflat
- Hong Kong China

2009 国际设计大奖
住宅设计 - 第一名
中国香港水蓝天岸16A样板房

Asia Pacific Residential Property Awards 2009
Best Interior Design Award
19A Le Bleu Deux Showflat
- Hong Kong China

2009 亚太住宅地产奖
最佳室内设计奖
中国香港水蓝天岸19A样板房

TID Taiwan Interior Design Award 2009
Residential Space Award
No.8 Tai Tam - Hong Kong China

2009 中国台湾室内设计奖
住宅空间奖
中国香港大潭道8号住宅

TID Taiwan Interior Design Award 2009
Residential Space Award
Parc Palais - Hong Kong China

2009 中国台湾室内设计奖
住宅空间奖
中国香港君颐峰住宅

Outstanding Greater China Design Awards 2009
Spatial Designer - Winner
No.8 Tai Tam - Hong Kong China

2009 大中华杰出设计大奖
空间设计奖
中国香港大潭道8号住宅

ELITE HOMES AWARDS 2009
The Most Outstanding House
No.8 Tai Tam - Hong Kong China

2009 精英住宅设计奖
最佳住宅奖
中国香港大潭道8号住宅

Perspective 40 Under 40
40 Outstanding Design Professionals
Under the Age of 40
Philip C.H. Tang & Brian S.H. Ip

40位40岁以下设计人才奖
邓子豪先生及叶绍雄先生

IF Design China Awards 2009
Gold Award
No.8 Tai Tam - Hong Kong China

2009 IF 中国设计大奖
金奖
中国香港大潭道8号住宅

China's Most Successful Design Awards 2009
Successful Design Award
Chelsea Residence - Shanghai, China
2009中国最成功设计奖
最成功设计
中国上海丁香御苑样板房

Guangzhou Design Week - Jinyang Prize 2009
Top 10 China Interior Design Team Award 2009
No.8 Tai Tam - Hong Kong China
2009广州设计周金羊奖
2009 中国十佳室内设计团队奖
中国香港大潭道8号住宅

The 7th Modern Decoration International Media Prize
Annual Residential Space Award
No.8 Tai Tam - Hong Kong China
第7届现代装饰国际传媒奖
年度住宅空间奖
中国香港大潭道8号住宅

G-MAGICCUBE AWARD 2009
Sample Unit - First Prize
The Sail at Victoria - Hong Kong China
G-MAGICCUBE AWARD 2009
样板房组 - 一等奖
中国香港傲翔湾畔样板房

G-MAGICCUBE AWARD 2009
Sample Unit - First Prize
Chelsea Residence Premier - Shanghai, China
G-MAGICCUBE AWARD 2009
样板房组 - 一等奖
中国上海丁香御苑样板房 Premier

G-MAGICCUBE AWARD 2009
Sample Unit - First Prize
Chelsea Residence Vista - Shanghai, China
G-MAGICCUBE AWARD 2009
样板房组 - 一等奖
中国上海丁香御苑样板房 Vista

G-MAGICCUBE AWARD 2009
Sample Unit - First Prize
16A Le Bleu Deux Showflat - Hong Kong China
G-MAGICCUBE AWARD 2009
样板房组 - 一等奖
中国香港水蓝天岸16A样板房

G-MAGICCUBE AWARD 2009
Sample Unit - Second Prize
Parc Palais - Hong Kong China
G-MAGICCUBE AWARD 2009
样板房组 - 二等奖
中国香港君颐峰住宅

G-MAGICCUBE AWARD 2009
Sample Unit - Second Prize
No.8 Tai Tam - Hong Kong China
G-MAGICCUBE AWARD 2009
样板房组 - 二等奖
中国香港大潭道8号住宅

**China Interior Design Awards
2009 (SK CUP)**
Residential - Third Prize
The Sail at Victoria   Hong Kong China
2009 中国室内设计大奖赛（尚高杯）
住宅组 - 三等奖
中国香港傲翔湾畔样板房

**China Interior Design Awards
2009 (SK CUP)**
Residential - Honorable Mention
No.8 Tai Tam - Hong Kong China
2009 中国室内设计大奖赛（尚高杯）
住宅组 - 荣誉奖
中国香港大潭道8号住宅

**China Interior Design Awards
2009 (SK CUP)**
Outstanding Award
Chelsea Residence Vista
- Shanghai, China
2009 中国室内设计大奖赛（尚高杯）
杰出奖
中国上海丁香御苑样板房 Vista

1st China's Top 50 Interior Design
Showflat Category - Winner
19A Le Bleu Deux Showflat
- Hong Kong China
第一届中国50佳室内设计
样板房组大奖
中国香港水蓝天岸19A样板房

1st China's Top 50 Interior Design
Showflat Category - Winner
16A Le Bleu Deux Showflat
- Hong Kong China
第一届中国50佳室内设计
样板房组大奖
中国香港水蓝天岸16A样板房

**CIID China Outstanding
Interior Designer 2009**
Winner
Philip C.H. Tang
2009 中国杰出室内设计师大奖赛
杰出室内设计师奖
邓子豪先生

**Awards
2009**

**Asia Pacific
Interior Design Awards 2008**
Honorable Award
Parc Palais - Hong Kong China

2008 亚太室内设计大奖
荣誉奖
中国香港君颐峰住宅

**Andrew Martin
Interior Design Awards 2008**
Top 50 International Interior Designers
Philip C.H. Tang & Brian S.H. Ip

2008 安德鲁·马丁室内设计大奖
全球50佳室内设计师奖
邓子豪先生及叶绍雄先生

IF Design China Awards 2008
Gold Award
Grand Promenade - Hong Kong China

2008 IF 中国设计大奖
金奖
中国香港嘉亨湾

International Design Awards 2008
Interior Room Products - Honorable Mention
Anglers' Bay Showflat - Hong Kong China

2008 国际设计大奖
室内设计专案 - 荣誉奖
中国香港海云轩样板房

International Design Awards 2008
Residential - Honorable Mention
Grand Promenade - Hong Kong China

2008 国际设计大奖
住宅组 - 荣誉奖
中国香港嘉亨湾

International Design Awards 2008
Residential - Honorable Mention
Parc Palais - Hong Kong China

2008 国际设计大奖
住宅组 - 荣誉奖
中国香港君颐峰住宅

**The 7th Macao Design
Biennial Awards 2008**
Outstanding Award
Anglers' Bay - Hong Kong China

2008 中国澳门第7届设计双年展
杰出奖
中国香港海云轩

**Asia Pacific
Interior Design Biennial Awards 2008**
Residential Unit - Gold Award
Grand Promenade - Hong Kong China

2008 亚太室内设计双年大奖赛
住宅组 - 金奖
中国香港嘉亨湾

**Asia Pacific
Interior Design Biennial Awards 2008**
Sample Unit - Gold Award
16A Le Bleu Deux Showflat
- Hong Kong China

2008 亚太室内设计双年大奖赛
样板房组 - 金奖
中国香港水蓝天岸16A样板房

**Asia Pacific
Interior Design Biennial Awards 2008**
Residential Unit - Bronze Award
Parc Palais - Hong Kong China

2008 亚太室内设计双年大奖赛
住宅组 - 铜奖
中国香港君颐峰住宅

**Asia Pacific
Interior Design Biennial Awards 2008**
Sample Unit - Bronze Award
La Rossa - Hong Kong China

2008 亚太室内设计双年大奖赛
样板房组 - 铜奖
中国香港影岸红样板房

**Asia Pacific
Interior Design Biennial Awards 2008**
Honorable Awards
Mangrove West Coast - China

2008 亚太室内设计双年大奖赛
荣誉奖
中国红树西岸样板房

**The 6th (2008) Modern Decoration
"International Media Prize"**
Annual Showflat Nomination Award
Mangrove West Coast - China

2008 第6届现代装饰国际传媒大奖
年度样板房提名奖
中国红树西岸样板房

**4th Chinese Interior Design
Awards 2008**
Gold Award
La Rossa - Hong Kong China

**2008 第4届海峡两岸
暨香港澳门室内设计大奖**
金奖
中国香港影岸红样板房

**4th Chinese Interior Design
Awards 2008**
Honorable Award
16A Le Bleu Deux Showflat
- Hong Kong China

**2008 第4届海峡两岸
暨香港澳门室内设计大奖**
荣誉奖
中国香港水蓝天岸16A样板房

**4th Chinese Interior Design
Awards 2008**
Honorable Award
Parc Palais - Hong Kong China

**2008 第4届海峡两岸
暨香港澳门室内设计大奖**
荣誉奖
中国香港君颐峰住宅

**Awards
2008**

4th Chinese Interior Design Awards 2008
Creative Award
19A Le Bleu Deux Showflat - Hong Kong China

2008 第4届海峡两岸暨香港澳门室内设计大奖
创意奖
中国香港水蓝天岸19A样板房

4th Chinese Interior Design Awards 2008
Creative Award
Grand Promenade - Hong Kong China

2008 第4届海峡两岸暨香港澳门室内设计大奖
创意奖
中国香港嘉亨湾

The RING iC@ward International Interior Design 2008
Gold Award
La Rossa - Hong Kong China

2008 金指环全球室内设计大奖
金奖
中国香港影岸红样板房

The RING iC@ward International Interior Design 2008
Silver Award
Grand Promenade - Hong Kong China

2008 金指环全球室内设计大奖
银奖
中国香港嘉亨湾

CIID China Interior Design Awards 2008
Best Showflat Space Design Award
La Rossa - Hong Kong China

2008 中国室内设计大奖赛
最佳样板房空间设计奖
中国香港影岸红样板房

CIID China Interior Design Awards 2008
Best Residence Space Design Award
Grand Promenade - Hong Kong China

2008 中国室内设计大奖赛
最佳住宅空间设计奖
中国香港嘉亨湾

China Interior Design Awards 2008 (SK CUP)
Silver Award
Parc Palais - Hong Kong China

2008 中国室内设计大奖赛（尚高杯）
银奖
中国香港君颐峰住宅

China Interior Design Awards 2008 (SK CUP)
Bronze Award
Grand Promenade - Hong Kong China

2008 中国室内设计大奖赛（尚高杯）
铜奖
中国香港嘉亨湾

China Interior Design Awards 2008 (SK CUP)
Honorable Award
19A Le Bleu Deux Showflat
- Hong Kong China

2008 中国室内设计大奖赛（尚高杯）
荣誉奖
中国香港水蓝天岸19A样板房

China Interior Design Awards 2008 (SK CUP)
Honorable Award
16A Le Bleu Deux Showflat
- Hong Kong China

2008 中国室内设计大奖赛（尚高杯）
荣誉奖
中国香港水蓝天岸16A样板房

CIID (Golden Beach) China Interior Design Awards 2008
Best Residence Award
16A Le Bleu Deux Showflat
- Hong Kong China

2008 中国国际建筑及室内设计节"金外滩"奖
最佳住宅奖
中国香港水蓝天岸16A样板房

Guangzhou Design Week - Jinyang Prize 2008
Residential Category
- Top 10 China Interior Design
16A Le Bleu Deux Showflat
- Hong Kong China

2008 广州设计周金羊奖
住宅组 - 中国十佳室内设计奖
中国香港水蓝天岸16A样板房

Guangzhou Design Week - Jinyang Prize 2008
Residential Category
- Top 10 China Interior Design
La Rossa - Hong Kong China

2008 广州设计周金羊奖
住宅组 - 中国十佳室内设计奖
中国香港影岸红样板房

Guangzhou Design Week - Jinyang Prize 2008
Villa Category - Top 10 China Interior Design
Parc Palais - Hong Kong China

2008 广州设计周金羊奖
别墅组 - 中国十佳室内设计奖
中国香港君颐峰住宅

The 7th China International Interior Design Biennial Awards 2008
Silver Award
16A Le Bleu Deux Showflat
- Hong Kong China

2008 第7届中国国际室内设计双年展
银奖
中国香港水蓝天岸16A样板房

The 7th China International Interior Design Biennial Awards 2008
Excellence Award
19A Le Bleu Deux Showflat
- Hong Kong China

2008 第7届中国国际室内设计双年展
优秀奖
中国香港水蓝天岸19A样板房

Awards 2008

The 7th China International
Interior Design Biennial Awards 2008
Excellence Award
La Rossa - Hong Kong China

2008 第7届中国国际室内设计双年展
优秀奖
中国香港影岸红样板房

The 7th China International
Interior Design Biennial Awards 2008
Excellence Award
Grand Promenade - Hong Kong China

2008 第7届中国国际室内设计双年展
优秀奖
中国香港嘉亨湾

PHOTO CREDITS

THANK YOU FOR THE SUPPORT FROM PHOTOGRAPHERS ALL THESE YEARS, VIRGILE SIMON BERTRAND AND ULSO TSANG.

ACKNOWLEDGEMENT

WE WOULD LIKE TO GIVE OUR SPECIAL THANK TO ALL THE PARTIES INVOLVED IN OUR PROJECTS AND THIS BOOK, INCLUDING OUR VALUABLE CLIENTS, CONTRACTORS, DESIGNERS, PHOTO-GRAPHERS, WRITERS AND PUBLISHER.

ROY CHAU, FANNY CHUNG, LINCOLN HO, CHERRIE LAI, BONNIE LAM, DOMINIC LAM, VIOLET LAM, MR. & MRS. LAW, APRIL LEE, TERRANCE LEE, JOEY NG, SANDY NG, WILLIAM PANG, VIVIAN SZE, MR. & MRS. TAM, SANDRA TAM, MR. WONG, ELVIS WONG, MR. & MRS. YAO, NOEL YUEN AND MR. & MRS. YUNG.

WITHOUT YOUR SUPPORT AND ASSISTANCE, WE WOULD NOT ABLE TO SHARE THESE EXCELLENT PROJECTS WITH READERS AROUND THE WORLD.